To my sister Audra
KWM

To Henry and Fitou the cat
SH

First published in Great Britain in 1995
This edition published 1997
Text copyright 1995 © Ken Wilson-Max
Illustration copyright 1995 © Sue Hendra
The moral right of the authors has been assured

Bloomsbury Publishing Plc, 38 Soho Square, London W1V 5DF
A CIP catalogue record for this book is available from
The British Library
ISBN 0 7475 30629
10 9 8 7 6 5 4 3 2 1
Manufactured in China

The Sun is a Bright Star

Ken Wilson-Max
Pictures by Sue Hendra

Bloomsbury Children's Books

On a clear night
look up at the deep, dark sky
and you'll see millions
of **stars**
- sharply twinkling.

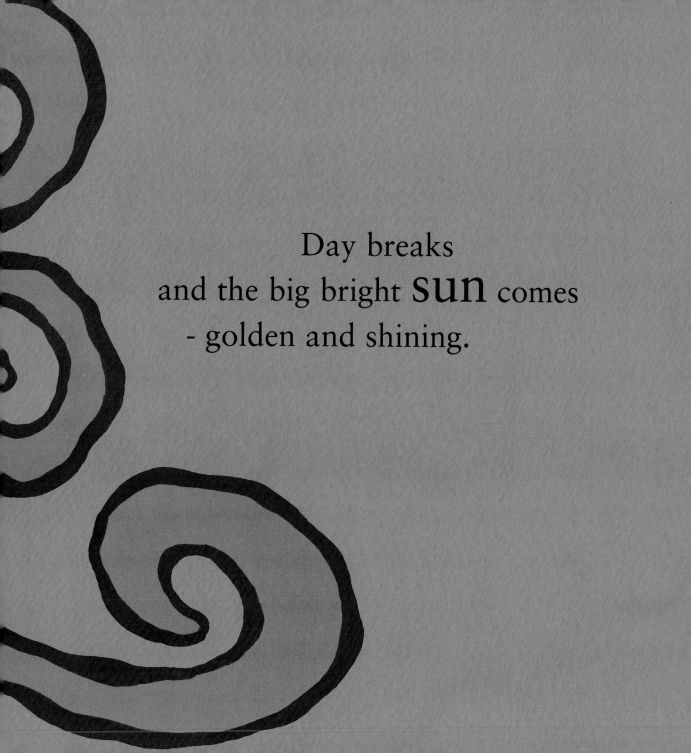

Day breaks
and the big bright sun comes
- golden and shining.

Mercury

Earth

Mars

Nine round
planets
circle the sun
- they never stop moving.

Venus

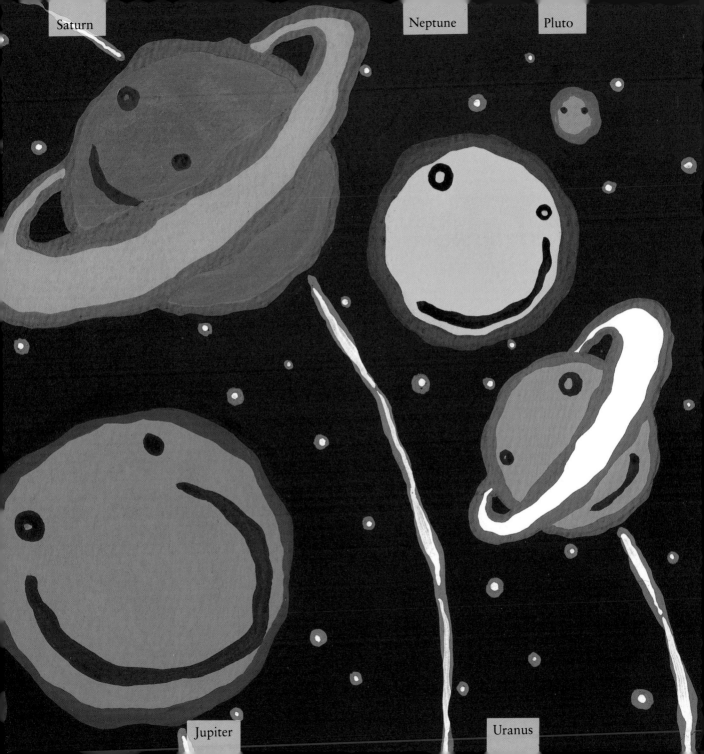

Saturn

Neptune

Pluto

Jupiter

Uranus

The little moon goes around the earth

Sometimes a **circle**,
a **half**,
or a **quarter**
- always beautiful.

Who
is looking
at us,
I wonder?